This Book Belongs To

Name : _____

Phone : _____

Address : _____

Project Name :

Foreman :

Project No :

Date :

Day :

Visitors	Schedule

Problems	Safety Issues

Summary Of Work

Signature :

Employee	Trade	Hours	Overtime

Equipment On Site	No. Units

Materials Delivered	No. Units	Equipment Rented	Rate

Others

Notes :

Project Name : _____

Foreman : _____

Project No :
Date :
Day :

Visitors	Schedule

Problems	Safety Issues

Summary Of Work

Signature : _____ ✏

Employee	Trade	Hours	Overtime

Equipment On Site	No. Units

Materials Delivered	No. Units	Equipment Rented	Rate

Others

Notes :

Project Name :

Foreman :

Project No :

Date :

Day :

Visitors

Schedule

Problems

Safety Issues

Summary Of Work

Signature :

Employee	Trade	Hours	Overtime

Equipment On Site	No. Units

Materials Delivered	No. Units	Equipment Rented	Rate

Others

Notes :

Project Name :

Foreman :

Project No :

Date :

Day :

Visitors	Schedule

Problems	Safety Issues

Summary Of Work

Signature :

Employee	Trade	Hours	Overtime

Equipment On Site	No. Units

Materials Delivered	No. Units	Equipment Rented	Rate

Others

Notes :

Project Name :

Foreman :

Project No :
Date :
Day :

Visitors

Schedule

Problems

Safety Issues

Summary Of Work

Signature :

Employee	Trade	Hours	Overtime

Equipment On Site	No. Units

Materials Delivered	No. Units	Equipment Rented	Rate

Others

Notes :

Project Name :

Foreman :

Project No :

Date :

Day :

Visitors

Schedule

Problems

Safety Issues

Summary Of Work

Signature :

Employee	Trade	Hours	Overtime

Equipment On Site	No. Units

Materials Delivered	No. Units	Equipment Rented	Rate

Others

Notes :

Project Name :

Foreman :

Project No :

Date :

Day :

Visitors

Schedule

Problems

Safety Issues

Summary Of Work

Signature :

Employee	Trade	Hours	Overtime

Equipment On Site	No. Units

Materials Delivered	No. Units	Equipment Rented	Rate

Others

Notes :

Project Name :

Foreman :

Project No :
Date :
Day :

Visitors

Schedule

Problems

Safety Issues

Summary Of Work

Signature :

Employee	Trade	Hours	Overtime

Equipment On Site	No. Units

Materials Delivered	No. Units	Equipment Rented	Rate

Others

Notes :

Project Name :

Foreman :

Project No :

Date :

Day :

Visitors

Schedule

Problems

Safety Issues

Summary Of Work

Signature :

Employee	Trade	Hours	Overtime

Equipment On Site	No. Units

Materials Delivered	No. Units	Equipment Rented	Rate

Others

Notes :

Project Name :	Project No :
	Date :
Foreman :	Day :

Visitors	Schedule

Problems	Safety Issues

Summary Of Work

Signature :

Employee	Trade	Hours	Overtime

Equipment On Site	No. Units

Materials Delivered	No. Units	Equipment Rented	Rate

Others

Notes :

Project Name :		Project No :
		Date :
Foreman :		Day :

Visitors	Schedule

Problems	Safety Issues

Summary Of Work

Signature :

Employee	Trade	Hours	Overtime

Equipment On Site	No. Units

Materials Delivered	No. Units	Equipment Rented	Rate

Others

Notes :

Project Name :

Foreman :

Project No :

Date :

Day :

Visitors

Schedule

Problems

Safety Issues

Summary Of Work

Signature :

Employee	Trade	Hours	Overtime

Equipment On Site	No. Units

Materials Delivered	No. Units	Equipment Rented	Rate

Others

Notes :

Project Name : _____

Foreman : _____

Project No :
Date :
Day :

Visitors	Schedule

Problems	Safety Issues

Summary Of Work

Signature : _____

Employee	Trade	Hours	Overtime

Equipment On Site	No. Units

Materials Delivered	No. Units	Equipment Rented	Rate

Others

Notes :

Project Name :

Foreman :

Project No :
Date :
Day :

Visitors

Schedule

Problems

Safety Issues

Summary Of Work

Signature :

Employee	Trade	Hours	Overtime

Equipment On Site	No. Units

Materials Delivered	No. Units	Equipment Rented	Rate

Others

Notes :

Project Name :

Foreman :

Project No :

Date :

Day :

Visitors

Schedule

Problems

Safety Issues

Summary Of Work

Signature :

Employee	Trade	Hours	Overtime

Equipment On Site	No. Units

Materials Delivered	No. Units	Equipment Rented	Rate

Others

Notes :

Project Name :

Foreman :

Project No :

Date :

Day :

Visitors	Schedule

Problems	Safety Issues

Summary Of Work

Signature :

Employee	Trade	Hours	Overtime

Equipment On Site	No. Units

Materials Delivered	No. Units	Equipment Rented	Rate

Others

Notes :

Project Name :

Foreman :

Project No :
Date :
Day :

Visitors	Schedule

Problems	Safety Issues

Summary Of Work

Signature :

Employee	Trade	Hours	Overtime

Equipment On Site	No. Units

Materials Delivered	No. Units	Equipment Rented	Rate

Others

Notes :

Project Name :

Foreman :

Project No :
Date :
Day :

Visitors	Schedule

Problems	Safety Issues

Summary Of Work

Signature :

Employee	Trade	Hours	Overtime

Equipment On Site	No. Units

Materials Delivered	No. Units	Equipment Rented	Rate

Others

Notes :

Project Name : _____

Foreman : _____

Project No :
Date :
Day :

Visitors

Schedule

Problems

Safety Issues

Summary Of Work

Signature : _____

Employee	Trade	Hours	Overtime

Equipment On Site	No. Units

Materials Delivered	No. Units	Equipment Rented	Rate

Others

Notes :

Project Name : _____

Foreman : _____

Project No :
Date :
Day :

Visitors	Schedule

Problems	Safety Issues

Summary Of Work

Signature : _____ ✏

Employee	Trade	Hours	Overtime

Equipment On Site	No. Units

Materials Delivered	No. Units	Equipment Rented	Rate

Others

Notes :

Project Name :

Foreman :

Project No :

Date :

Day :

Visitors

Schedule

Problems

Safety Issues

Summary Of Work

Signature :

Employee	Trade	Hours	Overtime

Equipment On Site	No. Units

Materials Delivered	No. Units	Equipment Rented	Rate

Others

Notes :

Project Name : _____

Foreman : _____

| Project No : |
| Date : |
| Day : |

Visitors	Schedule

Problems	Safety Issues

Summary Of Work

Signature : _____

Employee	Trade	Hours	Overtime

Equipment On Site	No. Units

Materials Delivered	No. Units	Equipment Rented	Rate

Others

Notes :

Project Name :

Foreman :

Project No :

Date :

Day :

Visitors

Schedule

Problems

Safety Issues

Summary Of Work

Signature :

Employee	Trade	Hours	Overtime

Equipment On Site	No. Units

Materials Delivered	No. Units	Equipment Rented	Rate

Others

Notes :

Project Name : _____

Foreman : _____

Project No :
Date :
Day :

Visitors	Schedule

Problems	Safety Issues

Summary Of Work

Signature : _____

Employee	Trade	Hours	Overtime

Equipment On Site	No. Units

Materials Delivered	No. Units	Equipment Rented	Rate

Others

Notes :

Project Name :

Foreman :

Project No :

Date :

Day :

Visitors

Schedule

Problems

Safety Issues

Summary Of Work

Signature :

Employee	Trade	Hours	Overtime

Equipment On Site	No. Units

Materials Delivered	No. Units	Equipment Rented	Rate

Others

Notes :

Project Name :

Foreman :

Project No :
Date :
Day :

Visitors	Schedule

Problems	Safety Issues

Summary Of Work

Signature :

Employee	Trade	Hours	Overtime

Equipment On Site	No. Units

Materials Delivered	No. Units	Equipment Rented	Rate

Others

Notes :

Project Name :

Foreman :

Project No :

Date :

Day :

Visitors	Schedule

Problems	Safety Issues

Summary Of Work

Signature :

Employee	Trade	Hours	Overtime

Equipment On Site	No. Units

Materials Delivered	No. Units	Equipment Rented	Rate

Others

Notes :

Project Name :

Foreman :

Project No :

Date :

Day :

Visitors

Schedule

Problems

Safety Issues

Summary Of Work

Signature :

Employee	Trade	Hours	Overtime

Equipment On Site	No. Units

Materials Delivered	No. Units	Equipment Rented	Rate

Others

Notes :

Project Name :

Foreman :

Project No :

Date :

Day :

Visitors	Schedule
Problems	Safety Issues

Summary Of Work

Signature :

Employee	Trade	Hours	Overtime

Equipment On Site	No. Units

Materials Delivered	No. Units	Equipment Rented	Rate

Others

Notes :

Project Name :

Foreman :

Project No :

Date :

Day :

Visitors

Schedule

Problems

Safety Issues

Summary Of Work

Signature :

Employee	Trade	Hours	Overtime

Equipment On Site	No. Units

Materials Delivered	No. Units	Equipment Rented	Rate

www.ingramcontent.com/pod-product-compliance
Lightning Source LLC
Chambersburg PA
CBHW050256120526
44590CB00016B/2377